YOUR KNOWLEDGE HAS VALUE

- We will publish your bachelor's and master's thesis, essays and papers

- Your own eBook and book - sold worldwide in all relevant shops

- Earn money with each sale

Upload your text at www.GRIN.com and publish for free

Vishal Vasistha

PID output fuzzified water level control in MIMO coupled tank system

GRIN Verlag

Bibliografische Information der Deutschen Nationalbibliothek:

Die Deutsche Bibliothek verzeichnet diese Publikation in der Deutschen National-
bibliografie; detaillierte bibliografische Daten sind im Internet über http://dnb.d-
nb.de/ abrufbar.

Imprint:

Copyright © 2013 GRIN Verlag GmbH
Druck und Bindung: Books on Demand GmbH, Norderstedt Germany
ISBN: 978-3-656-62813-2

GRIN - Your knowledge has value

Der GRIN Verlag publiziert seit 1998 wissenschaftliche Arbeiten von Studenten, Hochschullehrern und anderen Akademikern als eBook und gedrucktes Buch. Die Verlagswebsite www.grin.com ist die ideale Plattform zur Veröffentlichung von Hausarbeiten, Abschlussarbeiten, wissenschaftlichen Aufsätzen, Dissertationen und Fachbüchern.

Visit us on the internet:

http://www.grin.com/

http://www.facebook.com/grincom

http://www.twitter.com/grin_com

PID OUTPUT FUZZIFIED WATER LEVEL CONTROL IN MIMO COUPLED TANK SYSTEM

Vishal Vasistha

(Mechanical, National Institute of Technology Surathkal, India)

ABSTRACT: *The PID controllers are widely used in industry control applications due to their effectiveness and simplicity. This project presents PID controller design for MIMO coupled water tank level control system that is second order system. PID Controller output is fuzzified to control water level in coupled tank system. Simulation has been done in Matlab (Simulink library) with verification of mathematical model of controller. PID controller design and program has been prepared in LabVIEW. At the place of proportional valve, combinations of solenoid valves are used. The NI DAQ card is used for interfacing between hardware and LabVIEW software. Experiment is fully triggered by LabVIEW. Simulated results are compared with experimental results.*

Keywords – *PID, MIMO, Fuzzification, Coupled Tank, Control system*

1. INTRODUCTION

1.1 Overview

A lot of industrial applications of liquid level control are used now a day's such as in food processing, nuclear power generation plant, industrial chemical processing and pharmaceutical industries etc. The current work uses solenoid valves as actuators including of two small tanks mounted above a reservoir which functions as storage for the water. Each of both small tanks has independent pumps to pump water into the top of each tank. At the base of each tank, two flow valves (one as regular disturbance and other as leakage) connected to reservoir. In addition, capacitive-type probe level sensors have been used to monitor the level of water in each tank.

The PID Controller is usually used as temperature, motion and flow controllers. It is available in analog and digital forms. PID Controller controls the water flow rate through solenoid valves to maintain the required levels in both tanks. The NI DAQ card is used as the interface between hardware and software.

1

MATLAB 2012a (Simulink) has been used to get the simulation result of the system performance and LABVIEW 2010 to implement the designed controller. Fig. 1.1 shows the block diagram of the coupled tank control apparatus with controller.

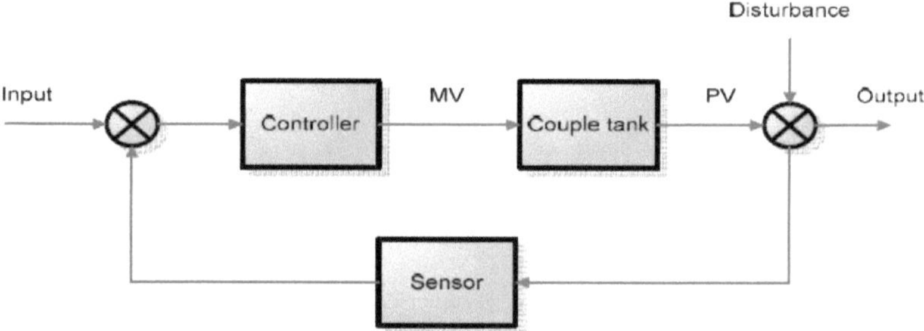

Figure 1.1: Block diagram of the couple-tank control apparatus.

1.2 Problem Statement

In spite of the simple structures, PID controllers are proven to be sufficient for many practical control problems. An abundant amount of research work has been reported in the past on the tuning of PID controllers. "PID" means Proportional-Integral-Derivative, referring to the three terms operating on the error signal to produce a control signal. Since many control systems using PID control have proved satisfactory, it still has a wide range of applications in industrial control.

In this project, several useful PID Controller design techniques are be presented, and implementation issues for the algorithms are also be discussed. The PID Controller was designed to control the liquid level in both the tanks. In this project, the simulation of proportional, integral and derivative actions are explained in detail, and variations of the basic PID structure are also introduced.

Finally, we need continuous data from the plant as the feedback, so to overcome this problem NI DAQ card have been used as the interfacing device between the hardware and software.

1.3 Objective

There are several objectives that must be achieved in order to make this project successful;

i.   To develop a PID Controller output fuzzified logic for controlling the liquid level in both the tanks of coupled tank system.

ii.   To validate the result from simulation (using MATLAB 2012a) through experimental set up (implementation using LabVIEW 2010).

1.4 Scope of project

This current work is all about how to designed the controller and simulate it using MATLAB 2012a. Then, implement PID Controller by developing GUI using LabVIEW 2010 software on coupled tank liquid level system. After that, both results are compared.

MATLAB 2012a has been used to simulate and verified the mathematical model of the controller. LabVIEW 2010 has been used to implement the graphical user interface for PID Controller.

The communication between DAQ card, LabVIEW 2010 and Coupled tank liquid level system has to be determined, in term of address to give or receive analog or digital signal.

1.5 Summary

This chapter is about the explanation for overall project. The objective and the scope of the project has to be given in order to give an insight about the idea of the project.

2. LITERATURE REVIEW

2.1 Overview

This chapter discusses the article those talks about different method of designing controllers.

2.2 Article

[1] Jutarut Chaorai-ngern, Arjin Numsomran, Taweepol Suesut, Thanit Trisuwannawat and Vittaya Tipsuwanporn.,2005, " PID Controller Design using Characteristic Ratio Assignment Method for Coupled-Tank Process",

This paper presents the PID controller design for coupled tank process using characteristic ratio assignment (CRA).The simulation results can be illustrated the validity of their approach by MATLAB.

[2] Muhammad Rehan, Fatima Tahir, Naeem Iqbal and Ghulam., 25-26 March 2008, " Modelling, Simulation and Decentralized Control of a Nonlinear Coupled Tank System",

This paper presents a coupled three tank system is taken as a plant and has been modelled mathematically using Bernoulli' s law, simulate with Matlab/Simulink and decentralized using control and estimation tool manager from Simulink model and mathematical model.

[3] M. Khalid Khan, Sarah K. Spurgeon, 10 February 2005 " Robust MIMO water level control in interconnected twin-tanks using second order sliding mode control",

This paper presents a coupled twin-tank system as a plant and water level is controlled by second order sliding mode control and simulation by Matlab Simulink.

[4] Liu Jinkun., 2004, MATLAB simulation of advanced PID control.

This paper gives detailed explanation about Matlab simulation of PID controller.

2.3 Summary

This chapter is about the explanation for some articles that refer to get the information or some knowledge that applies to make the project run successfully. There are several articles that explain about same controller that has been used in this project but applied at different plant or different method of designing, and also different controller but applied at same plant.

3. METHODOLOGY

3.1 Overview

This chapter discusses about the method that has been used to complete this project. The coupled twin-tank liquid level system has been used in this project as a plant. In this project, several useful PID Controller design techniques are presented. The PID Controller and PID controller output fuzzification has been designed to control the liquid level in both tanks to obtain the desired level. In this current work, NI DAQ card has been used as the interfacing between the hardware and software.

3.2 Project Flow Chart

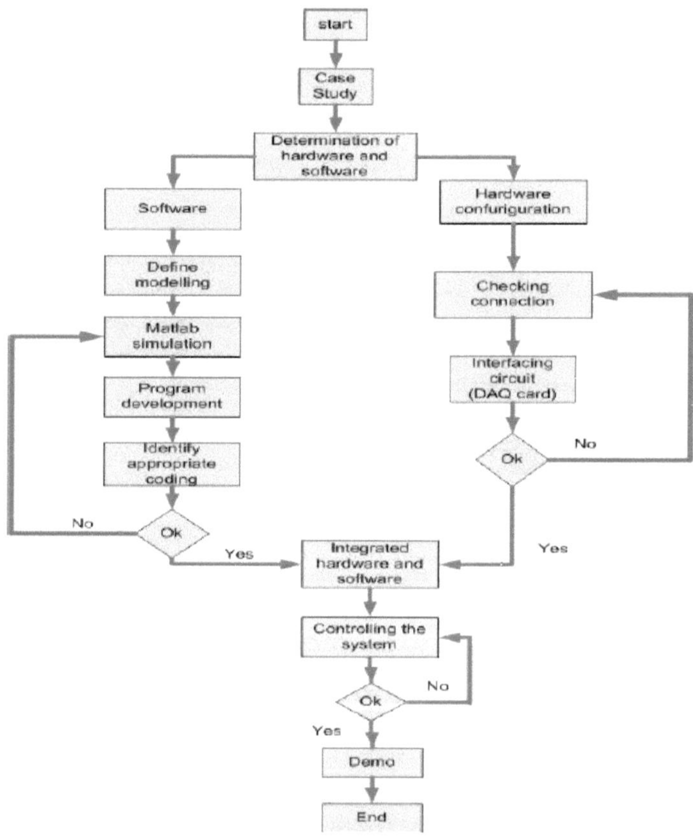

Figure 3.1: Flow chart for software and hardware development

Fig. 3.1 shows about the overall progress for both software and hardware development that has been discussed later. This project has been divided to two parts to make sure this project runs smoothly. The first part is a software part, which covers modelling the controller. The controller is designed and simulated using MATLAB 2012a and then implemented in LabVIEW 2010 as GUI with fuzzified logic. The second part covers for the hardware part. In this part, the coupled tank liquid level system has been assembled to make sure it runs properly. Then, a communication between plant and controller has been made using NI DAQ card. The DAQ card is analyzed and these operations need to refer the manual to make the interfacing between plant and controller.

After that, both parts are integrated to test the whole system. At this part, the results that implemented using LabVIEW 2010 is compared with the experiment result that simulated using MATLAB 2012a. Troubleshooting has been performed to obtain better results in case of some error occurs.

3.3 Mathematical Modelling of Coupled Tank System

Before the process of designing controller begin, it is vital to understand the mathematics of how the coupled tank system behaves. In this system, nonlinearity in the dynamic model has been observed.

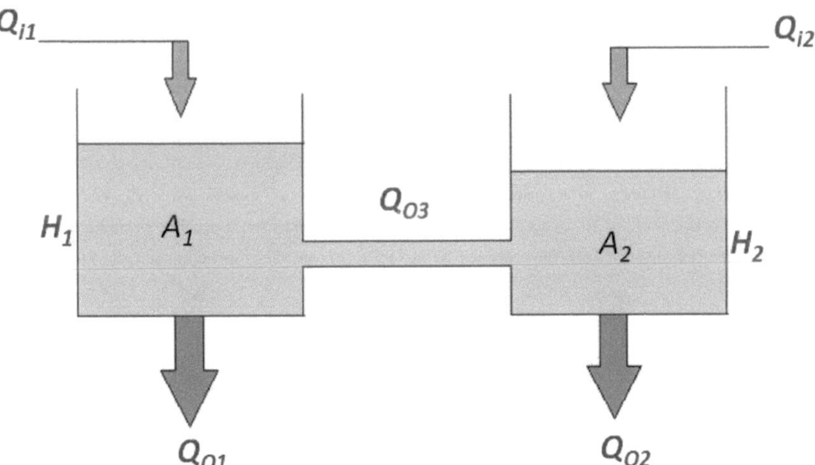

Figure 3.2: Schematic diagram of coupled tank system

3.3.1 A Simple Nonlinear Model of Coupled Tank System

A simple nonlinear model is derived based on figure 3.2. Let H1and H2 be the fluid level in each tank, measured with respect to the corresponding outlet. Considering a simple mass balance, the rate of change of fluid volume in each tank equals the net flow of fluid into the tank. Thus for each of tank 1 and tank 2, the dynamic equation is developed as follows:

$$A_1 = \left(\frac{dH_1}{dt}\right) Q_{i1} - Q_{o1} - Q_{o3} \qquad \ldots\ldots (3.3.1.1)$$

$$A_2 = \left(\frac{dH_2}{dt}\right) Q_{i2} - Q_{o2} + Q_{o3} \qquad \ldots\ldots (3.3.1.2)$$

Where

H_1, H_2 = height of fluid in tank 1 and tank 2 respectively

A_1, A_2 = cross sectional area of tank 1 and tank 2 respectively

Q_{o3} = flow rate of fluid between tanks

Q_{i1}, Q_{i2} = pump flow rate into tank 1 and tank 2 respectively

Q_{o1}, Q_{o2} = flow rate of fluid out of tank 1 and tank 2 respectively

Each outlet drain can be modelled as a simple orifice. Bernoulli's equation for steady, non viscous, incompressible shows that the outlet flows in each tank is proportional to the square root of the head of water in the tank. Similarly, the flow between the two tanks is proportional to the square root of the head differential.

$$Q_{O1} = \alpha_1 \sqrt{H_1} \qquad \ldots\ldots (3.3.1.3)$$

$$Q_{O2} = \alpha_2 \sqrt{H_2} \qquad \ldots\ldots (3.3.1.4)$$

$$Q_{O3} = \alpha_3 \sqrt{(H_1 - H_2)} \qquad \ldots\ldots (3.3.1.5)$$

Where $\alpha_1, \alpha_2, \alpha_3$ are proportional constants which depend on the coefficients of discharge, the cross sectional area of each orifice and the gravitational constant.

Combining equation (3.3.1.3), (3.3.1.4) and (3.3.1.5) into equations (3.3.1.1) and (3.3.1.2), a set of nonlinear state equations which describe the system dynamics of the coupled tank are derived.

$$A_1 \left(\frac{dH_1}{dt} \right) = Q_{i1} - \alpha_1 \sqrt{H_1} - \alpha_3 \sqrt{(H_1 - H_2)} \qquad \ldots\ldots (3.3.1.6)$$

$$A_2 \left(\frac{dH_2}{dt} \right) = Q_{i2} - \alpha_2 \sqrt{H_2} + \alpha_3 \sqrt{(H_1 - H_2)} \qquad \ldots\ldots (3.3.1.7)$$

3.3.2 A Linearised Perturbation Model

For a set of inflows Q_{i1} and Q_{i2}, the fluid level in the tanks is at some steady state level H_1 and H_2. Consider a small variation in each inflow, q_1 in Q_{i1} and q_2 in Q_{i2}. Let the resulting perturbation in level be $h1$ and $h2$ respectively. From equations (3.3.1.6) and (3.3.1.7), the equation becomes:

For Tank 1

$$A_1 \frac{d(H_1 + h1)}{dt} = (Q_{i1} + q_1) - \alpha_1 \sqrt{(H_1 + h_1)} - \alpha_3 \sqrt{(H_1 - H_2)} + (h_1 - h_2)$$

$$\ldots\ldots (3.3.2.1)$$

For Tank 2

$$A_2 \frac{d(H_2 + h_2)}{dt} = (Q_{i2} + q_2) - \alpha_2 \sqrt{(H_2 + h_2)} + \alpha_3 \sqrt{(H_1 - H_2)} + (h_1 - h_2)$$

$$\ldots\ldots (3.3.2.2)$$

Subtracting equations (3.3.1.6) and (3.3.1.7) from equation (3.3.2.1) and (3.3.2.2), the equations obtained are,

$$A_1 \left(\frac{dh_1}{dt} \right) = q_1 - \alpha_1 \{ \sqrt{(H_1 + h_1)} - \sqrt{H_1} \} - \alpha_3 \{ \sqrt{(H_1 - H_2 + h_1 - h_2)} - \sqrt{(H_1 - H_2)} \}$$

$$\ldots\ldots (3.3.2.3)$$

$$A_2\left(\frac{dh_2}{dt}\right) = q_2 - \alpha_2\{\sqrt{(H_2 + h_2)} - \sqrt{H_2}\} + \alpha_3\{\sqrt{(H_1 - H_2 + h_1 - h_2)} - \sqrt{(H_1 - H_2)}\}$$

$$\ldots\ldots (3.3.2.4)$$

For small perturbations,

$$\sqrt{(H_1 + h_1)} = \sqrt{H_1\left(1 + \frac{H_1}{2H_1}\right)} \qquad \ldots\ldots (3.3.2.5)$$

Therefore,

$$\sqrt{(H_1 + h_1)} - \sqrt{H_1} \approx \frac{h_1}{2\sqrt{H_1}}$$

Similarly,

$$\sqrt{(H_2 + h_2)} - \sqrt{H_2} \approx \frac{h_2}{2\sqrt{H_2}}$$

And

$$\sqrt{(H_2 - H_1 + h_2 - h_1)} - \sqrt{H_2 - H_1} \approx \frac{h_2 - h_1}{2\sqrt{(H_2 - H_1)}}$$

Simplify equation (3.3.2.3) and (3.3.2.4) with these approximations becomes,

$$A_1\left(\frac{dh_1}{dt}\right) = q_1 - \frac{\alpha_1}{2\sqrt{H_1}}h_1 - \{\alpha_3/2\sqrt{(H_1 - H_2)}\}(h_1 - h_2) \qquad \ldots\ldots (3.3.2.6)$$

$$A_2\left(\frac{dh_2}{dt}\right) = q_2 - \frac{\alpha_2}{2\sqrt{H_2}}h_2 + \{\alpha_3/2\sqrt{(H_1 - H_2)}\}(h_1 - h_2) \qquad \ldots\ldots (3.3.2.7)$$

In equations (3.3.2.6) and (3.3.2.7), note that the coefficients of the perturbations in level are functions of the steady state operating points H_1 and H_2. Note that the two equations can also be written in the form

$$A_1\left(\frac{dh_1}{dt}\right) = q_1 - q_{o1} - \{\alpha_3/2\sqrt{(H_1 - H_2)}\}(h_1 - h_2) \qquad \ldots\ldots (3.3.2.8)$$

$$A_2\left(\frac{dh_2}{dt}\right) = q_2 - q_{o2} + \{\alpha_3/2\sqrt{(H_1 - H_2)}\}(h_1 - h_2) \qquad \ldots\ldots (3.3.2.9)$$

Where q_{o1} and q_{o2} represent perturbations in the outflow at the drain pipes. This is appropriate in the case where outflow is controlled by attaching an external clamp for instance.

3.4 System or plant Design

Each value of $\alpha_1, \alpha_2, \alpha_3, A_1, A_2, H_1$ and H_2 can be obtained from mathematical modelling equations-

Table 3.1: Parameters values

$H_1 = 20$
$H_2 = 17$
$\alpha_1 = 53.436$
$\alpha_2 = 53.436$
$\alpha_3 = 53.436$
$A_1 = 600$
$A_2 = 600$

By using the Parameters value and equations (3.3.2.6), (3.3.2.7), we can get the following equations in the form of manipulating variables q_1, q_2 and process variables h_1, h_2 -

$$\frac{dh_1}{dt} = 1.67e(-3)\, q_1 - 9.96e(-3)\, h_1 - 0.0257\,(h_1 - h_2) \qquad \dots (3.4.1)$$

$$\frac{dh_2}{dt} = 1.67e(-3)\, q_2 - 0.0108\, h_2 + 0.0257(h_1 - h_2) \qquad \dots (3.4.2)$$

$$\begin{pmatrix} \dot{h_1} \\ \dot{h_2} \end{pmatrix} = \begin{bmatrix} -0.03566 & 0.0257 \\ 0.0257 & -0.0356 \end{bmatrix} \begin{pmatrix} h_1 \\ h_2 \end{pmatrix} + \begin{bmatrix} 1.67e - 3 & 0 \\ 0 & 1.67e - 3 \end{bmatrix} \begin{pmatrix} q_1 \\ q_2 \end{pmatrix}$$

$$\begin{pmatrix} y_1 \\ y_2 \end{pmatrix} = \begin{bmatrix} 1 & 0 \\ 0 & 1 \end{bmatrix} \begin{pmatrix} h_1 \\ h_2 \end{pmatrix} + \begin{bmatrix} 0 & 0 \\ 0 & 0 \end{bmatrix} \begin{pmatrix} q_1 \\ q_2 \end{pmatrix} \qquad\qquad ..... (3.4.3)$$

Above equation (3.4.3) is the transfer function of coupled tank system in the form of state-space matrices, where

$\dot{h_1}$ = derivative of state variable for tank 1

$\dot{h_2}$ = derivative of state variable for tank 2

h_1 = state variable for tank 1

h_2 = state variable for tank 2

q_1 = input variable for tank 1

q_2 = input variable for tank 2

y_1 = output variable for tank 1

y_2 = output variable for tank 2

3.5 MATLAB R2012a

This topic presents the designing of PID Controller to control coupled tank system using MATLAB R2012a software. This software is used to create the Simulink diagram for PID Controller and performance for each parameter for PID Controller is also simulated. The performances of PID Controller are evaluated in terms of overshoot, rise time and steady state error. Then, the gain for each parameter also has been tuned in this software and the validity for each parameter is compared using the reference value (set point). Fig. 3.3 shows the MATLAB Simulink block for PID Controller combines with plant.

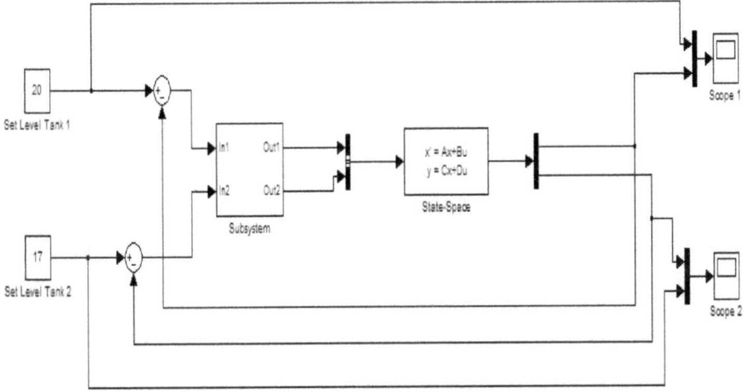

Figure 3.3: Block Diagram of PID Controller combines with plant

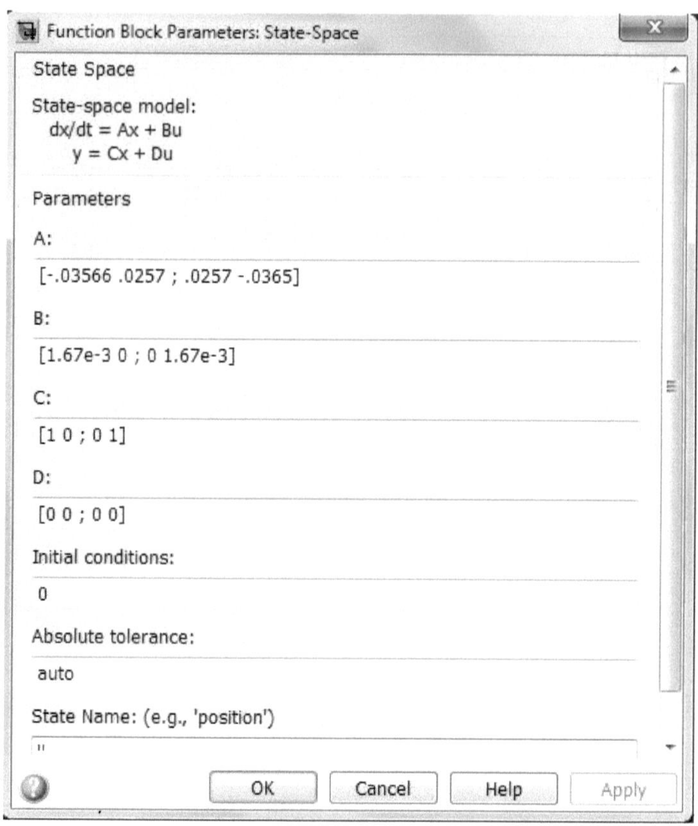

Figure 3.4: State-space matrices values in Matlab

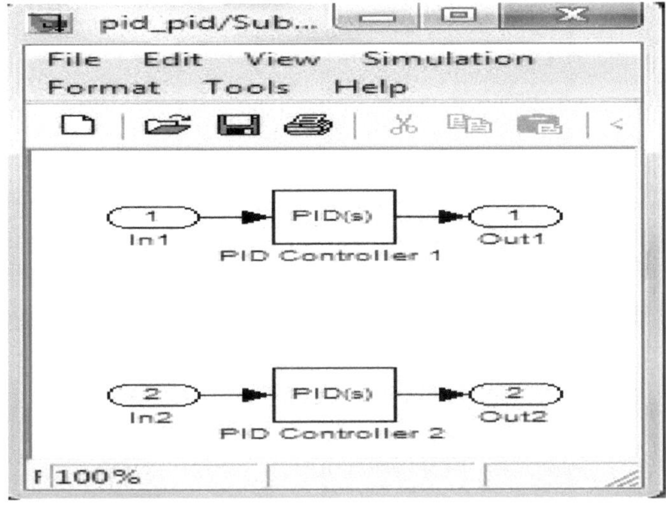

Figure 3.5: PID Subsystem components in Matlab

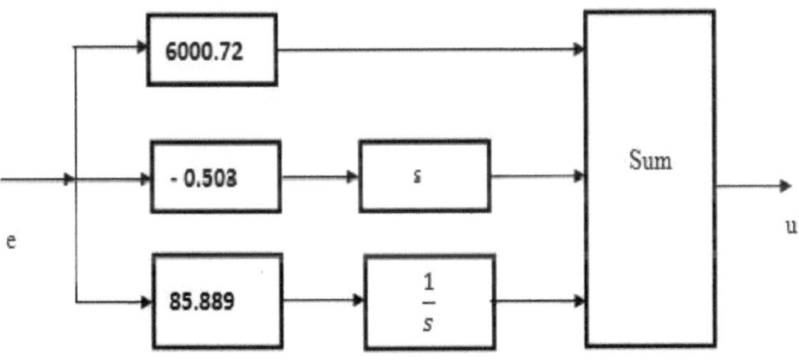

Figure 3.6: Block Diagram of inside PID Controller 1

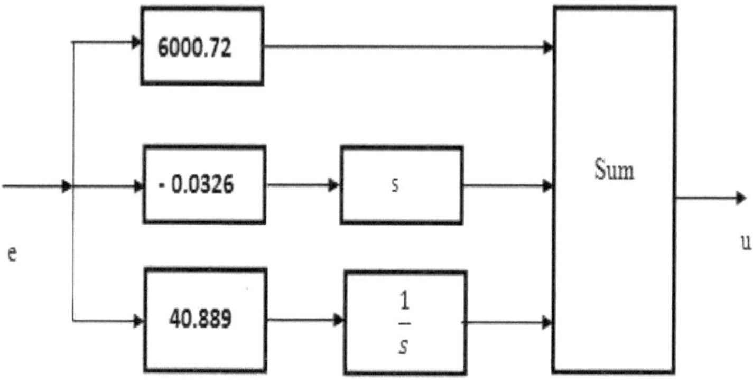

Figure 3.7: Block Diagram of inside PID Controller 2

Based on the transfer function for equation (3.4.3), state-space matrices values are fed in Matlab, shown in fig. 3.4. Fig. 3.6 and 3.7 are the controllers for both the tanks in this system. This controller is design based on equation of PID Controller,

$$u(t) = K_p\, e(t) + K_i \int e(t)dt + K_d\, \frac{de}{dt}(t)$$

$$u(t) = K_p\, [e(t) + \frac{1}{T_i}\int e(t)dt + T_d\, \frac{de}{dt}(t)]$$

$$..... (3.5.1)$$

Where $K_i = K_p / T_i :$

              $K_d = K_p . T_d$

3.6 LabVIEW 2010

This topic presents the designing of PID Controller to control coupled tank system using LabVIEW 2010 software. This software is used for getting the implement result for the project by develop a GUI for PID Controller. Before the GUI for PID Controller is programmed, algorithm for PID Controller is needed. So that, to find the algorithm the set point is compared to the process variable to obtain the error.

Error = SP– PV

Then, convert equation (3.5.1) becomes,

$$u(s) = K_p \left(1 + \frac{1}{T_i\, s} + T_d\, s\right)$$

$$u = K_p \left[error + \frac{error\ new - error\ old}{T_i} + (error\ new - error\ old) * T_d\right]$$

$$..... (3.6.1)$$

The GUI for controller is created in LabVIEW 2010.

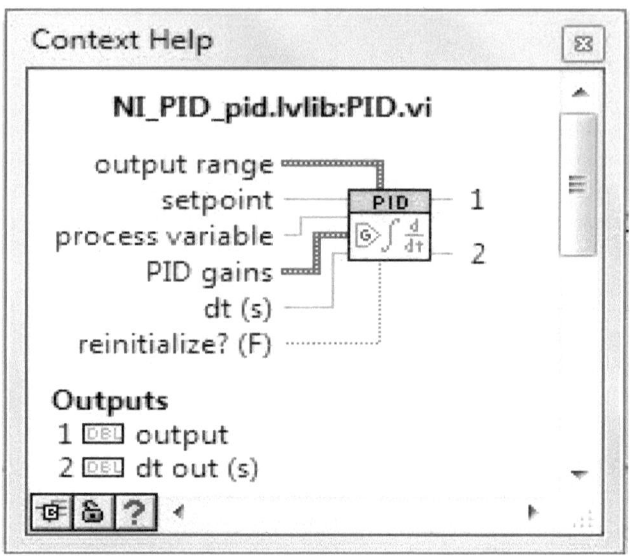

Figure 3.8: PID Controller in LabVIEW

Above fig. 3.8 shows input and outputs PID controller used in LabVIEW 2010 as GUI.

The detailed explanation of the each input and output shown below-

- Output Range specifies the range to which coerce the control output.

- Set point is the desired value of the process variable to control.

- Process variable specifies the measured value of the process variable being controlled. This value is equal to the feedback value of the feedback control loop.

- PID gains specify the proportional gain, integral time, and derivative time parameters of the controller. The parameters are defined in terms of proportional gain (Kc), integral time (Ti, min), derivative time (Td, min).

18

- dt (s) - specifies the interval, in seconds, at which this VI is called.

- Reinitialize specifies whether to reinitialize the internal parameters, such as the integrated error, of the controller.

- Output returns the control output of the PID algorithm that is applied to the controlled process.

- dt out (s) returns the actual time interval in seconds.

First created GUI is-

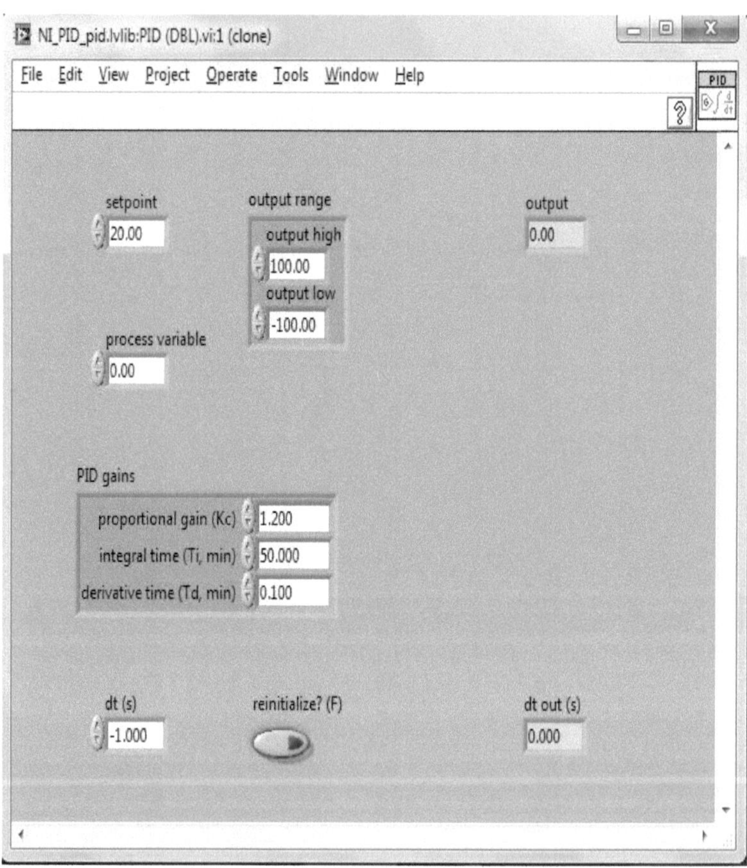

Figure 3.9: The GUI for PID Controller 1

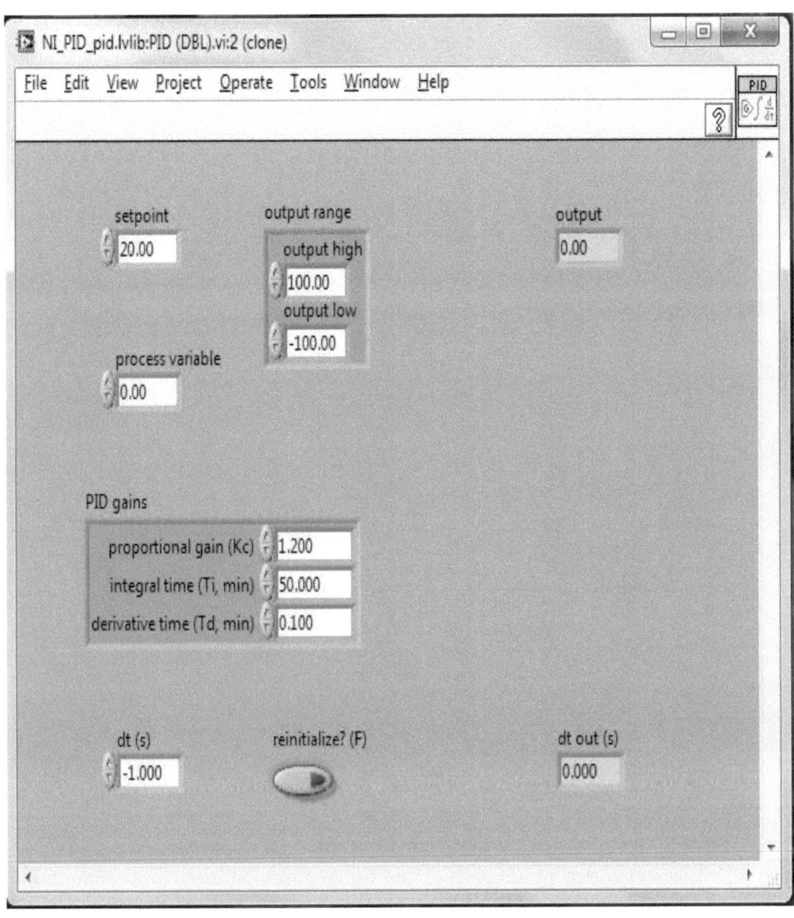

Figure 3.10: The GUI for PID Controller 2

Fig. 3.9 and 3.10 show the first GUI that has been created. This GUI is created based on the algorithm for PID Controller that had been stated on equation (3.6.1).

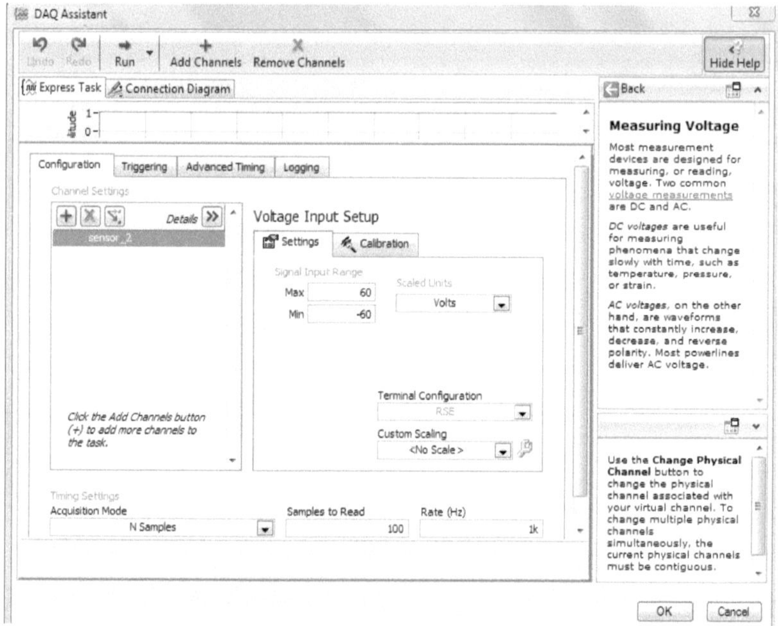

Figure 3.11: The GUI for DAQ card analog input detection

Fig. 3.11 shows the second GUI for any one tank sensor analog input measurement that been created. GUI for another tank sensor is same as first one. NI module 9221 is used for analog input measurement. This GUI is the first GUI that runs once the program is started. This GUI is created to detect the DAQ card that been used. The Run button appears after GUI detects the DAQ card and clicked to proceed to next step.

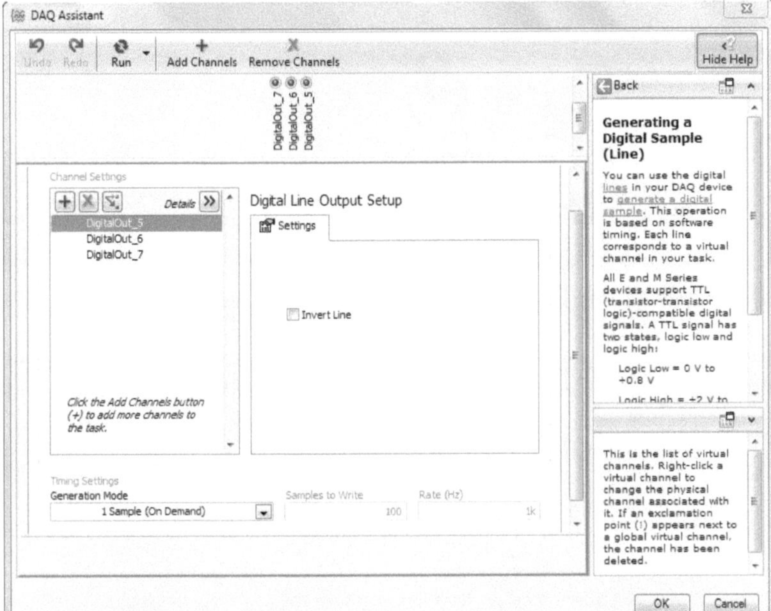

Figure 3.12: The GUI for DAQ card digital output detection

Fig 3.12 shows the third GUI for any one tank digital output measurement that has been created. GUI for another tank digital output measurement be same as first one. NI modules 9474 and 9472 have been used for digital output measurement. This GUI is created to detect the DAQ card that been used. The Run button appears after GUI detects the DAQ card and clicked to proceed to next step.

3.7 DAQ Card

The NI cDAQ with NI A/I and A/O Modules has been used as the data acquisition input output card for the experimental implementation. Fig. 3.13 shows the DAQ card functions to communicate between controller and plant.

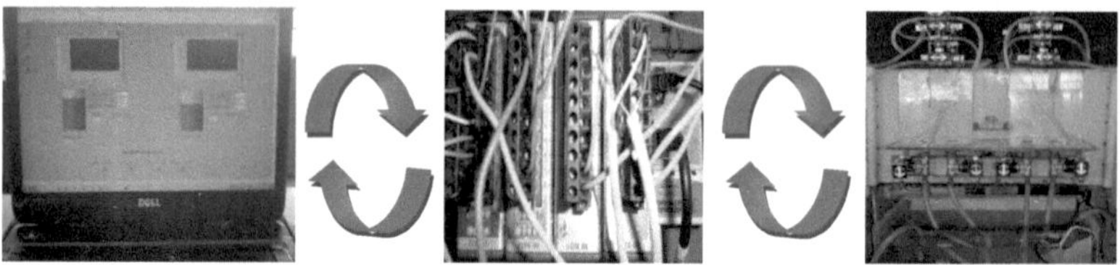

Figure 3.13: NI DAQ card connection

The designed controllers are sending the required signal to solenoid valves at the coupled tanks. These signals must flow through the DAQ card. Then, the DAQ card sends these signals to solenoid valves in the coupled tanks system. Each coupled tank which consists of sensor and actuator is in a continuos closed loop to send back the signal to the controller for next iteration.

3.8 Summary

This section gave an overview about the entire necessary thing that involved in this project. The coupled tank, software and hardware related also had been discussed in terms of it characteristics and it important parameter.

4. RESULT, ANALYSIS AND DISCUSSION

4.1 Overview

This chapter presents the simulation and implementation result using PID Controller for the coupled tank system. The performances for each parameter of PID Controller are evaluated based on rise time, overshoot and steady state error. The comparison between two results is also discussed in this chapter.

4.2 Simulation results for different type of controller used for tank 1

This section shows the simulation result for different type of controller used for tank- 1 water level control. The equation for coupled tank system refers the equation (3.4.3). Fig 4.1 shows tank 1 level control for different controllers response comparison.

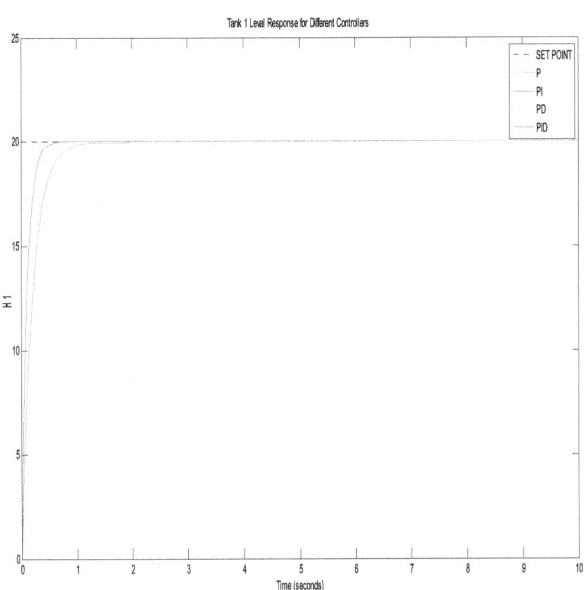

Figure 4.1: Tank 1 level response for different controllers

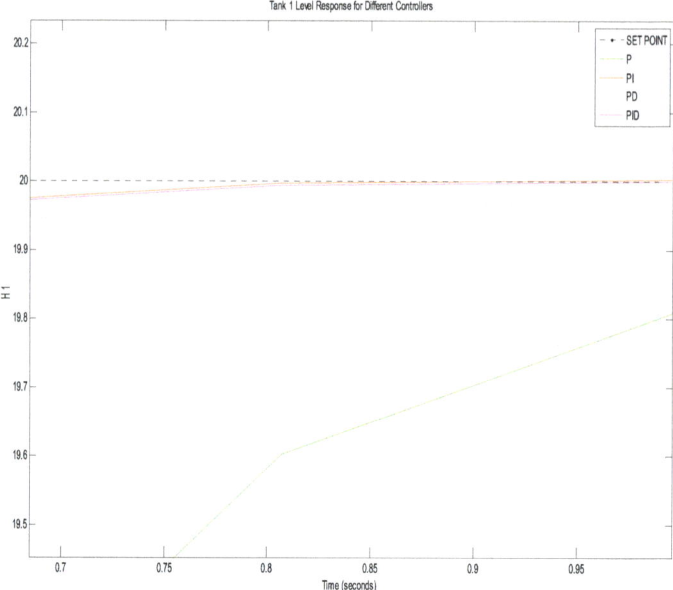

Figure 4.2: Tank 1 level response for different controllers in terms of rise time and overshoot/undershoot

From fig. 4.2 it is clear that for P controller rise time is more compare to other controllers. PI controller has overshoot while P and PD controllers give undershoot. Comparing with other controllers, PID has less rise time and more stable (no overshoot/undershoot), so PID controller is most effective than any other controller.

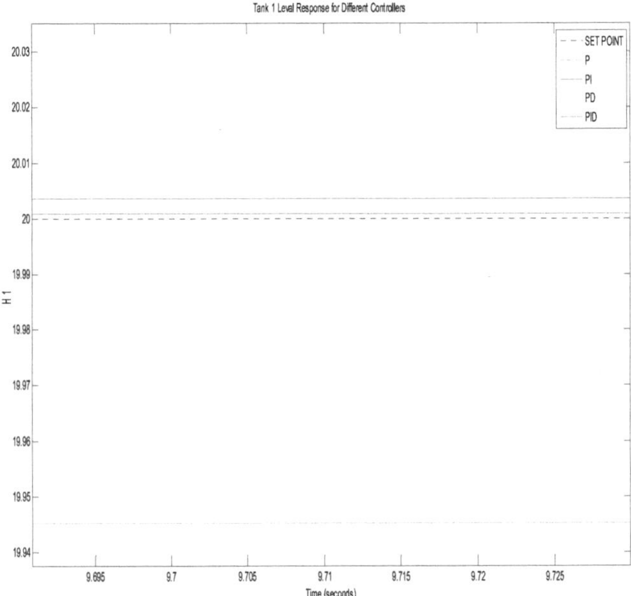

Figure 4.3: Tank 1 level response for different controllers in terms steady state error (S.S.E.)

From fig. 4.3 it is shown that for P controller has more steady state error (S.S.E.) compare to other controllers. PD controller gives moderate SSE. PI controller gives little SSE but more than PID controller.PID has least SSE and more accurate than any other controller.

4.3 Simulation results for different type of controller used for tank 2

This section shows the simulation result for different type of controllers used for tank-2 water level control. The equation for coupled tank system refers the equation (3.4.3) as same as for tank-1. Fig. 4.1 shows tank 2 level control for different controllers response comparison.

27

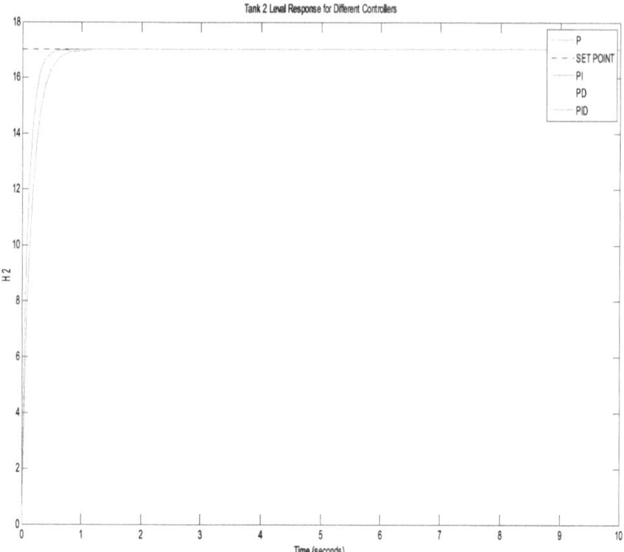

Figure 4.4: Tank-2 level response for different controllers

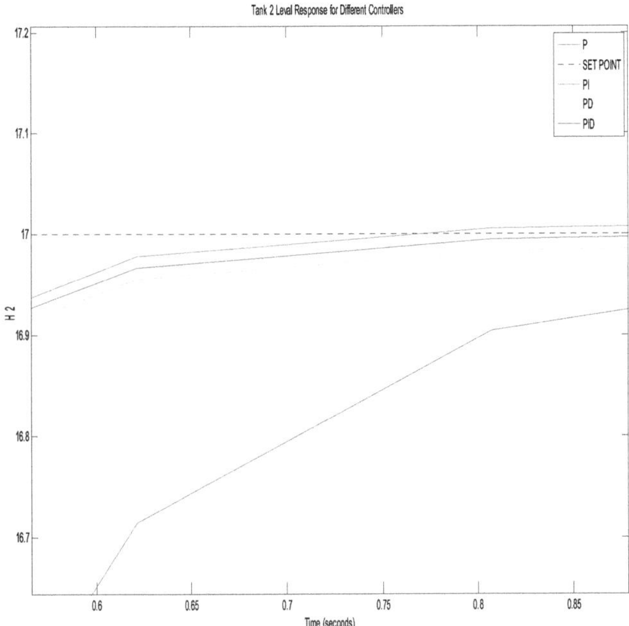

Figure 4.5: Tank 1 level response for different controllers in terms of rise time and overshoot/undershoot

From fig. 4.5 it is clear that for P controller rise time is more compare to other controllers. PI controller has overshoot while P and PD controllers give undershoot. Comparing with other controllers, PID has less rise time and more stable (no overshoot/undershoot), so PID controller is most effective than any other controller.

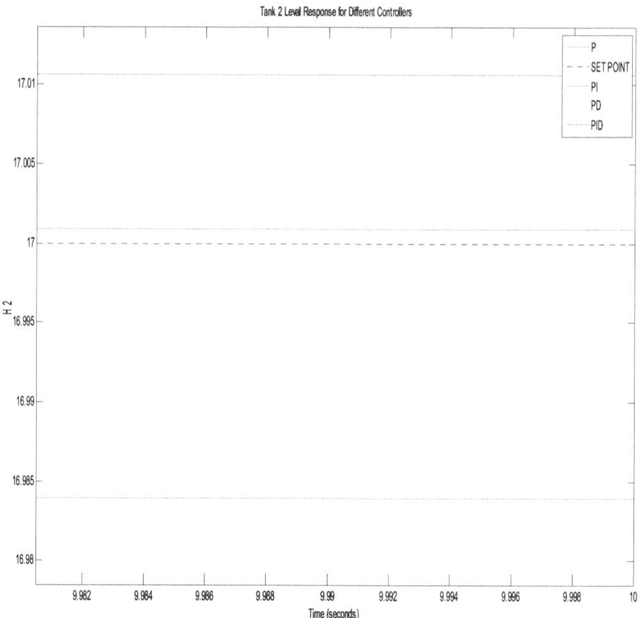

Figure 4.6: Tank 2 level response for different controllers in terms steady state error (S.S.E.)

From fig. 4.6 it is shown that for P controller has more steady state error (S.S.E.) compare to other controllers. PD and PI controller gives moderate SSE but more than PID controller. PID has least SSE and more accurate than any other controller.

4.4 Experimental Result

This section shows the experimental result as the PID Controller output fuzzified logic controls liquid level at tank 1 and tank 2. The configuration of the GUI model is shown at section 3.6. The performance result for level liquid that is controlled by PID Controller has been discussed.

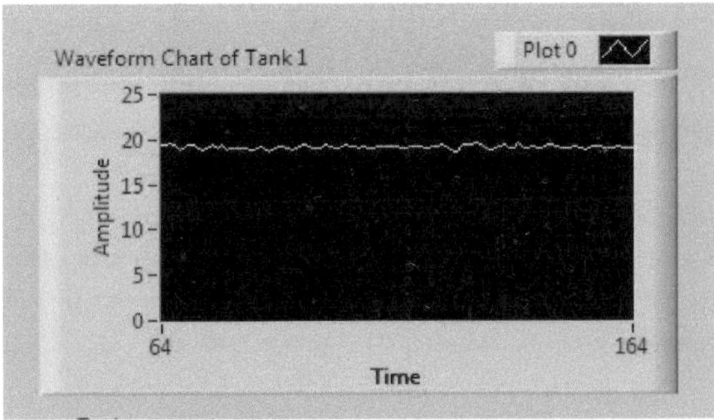

Figure 4.7: Tank-1 maintained level using PID Controller

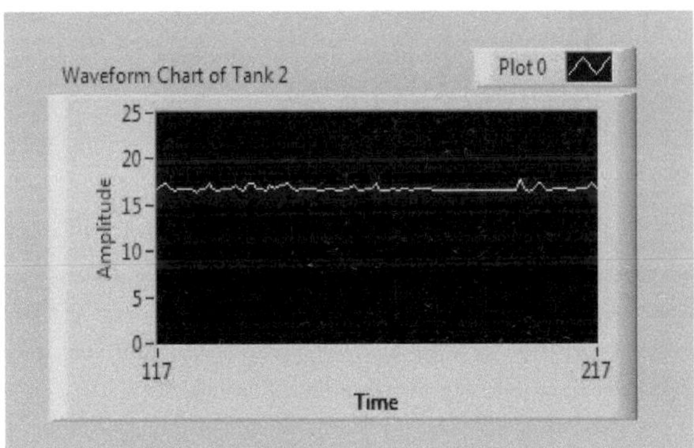

Figure 4.8: Tank-2 maintained level using PID Controller

Fig. 4.7 and 4.8 shows the result when PID Controller is controlling water level in tank-1and tank-2 at coupled tank system. The set points (20 cm for tank-1 and 17 cm for tank-2) are set for both the tanks. The proportional gain is set equal to 1.2, integral time is set equal to 50 min and derivative time is set equal to 0.1 min to provide the desired response. After the Start/Run button is clicking, the controller starts to run and send desired voltage to solenoid valves combinations at tank-1 and tank-2. The value of desired voltage controls manipulating variable (flow rate) through solenoid valves combinations. As level is going to be maintained in both the tanks, it is shown on waveform chart at front panel.

4.5 Discussion for PID Controller

Because of the action of Proportional parameter, the plot result responds to a change very quickly. Due to the action of Integral parameter, the system is able to be returned to the set point value. The Derivative parameter measures the change in the error and help to adjust the plot result accordingly. Table 4.1 shows the effects of increasing proportional, integral and derivative parameters.

Table 4.1: Effects of increasing parameter

Parameter	Rise Time	Overshoot	Settling Time	SSE
K_p	decreased	increase	Small change	decreased
K_i	decreased	increase	increase	eliminate
K_d	Small change	decreased	decreased	none

4.6 Comparison between simulation and implementation result

The objective of comparing the result of PID Controller that control liquid level at both the tanks on coupled tank between the simulation and implementation result is to investigate to find the better result of PID Controller. Design techniques of simulation and implementation have been explored and their performance is evaluated base on percentage overshoot, settling time and steady state error.

It is shown that, the simulation result achieve the set point voltage as show in fig. 4.1 and 4.4. The simulation result showed the steady state error value is nearly 0%. The settling time is the time for response to reach and stay within the set point and for simulation result is very less around 1 second. The simulation result does not have any percentage overshoot.

Therefore, the implementation result does not achieve the set point exactly as it required. As it shows in fig. 4.7 and 4.8, the steady state error value exists there, even it's very small. In the implementation result, there are no percentage overshoot but some settling time because the plot does not achieve the set point exactly.

4.7 Summary

This chapter discussed the result obtained for both simulation and implementation result of PID Controller. The result had been compared, and simulation result shows the better result than real time result.

5. CONCLUSION AND FUTURE RECOMMENDATION

5.1 Conclusion

As a conclusion, PID Controller had been successfully designed to control liquid level at both the tanks on coupled tank system using simulation and implementation. The comparison has been made and simulation techniques perform better result as compared to the implementation.

The advantage of simulation technique is that using block diagram is easy to run and execute the program. Therefore, there is no need to find the algorithm for PID Controller. There are some difficulties for implementation technique due to the hardware involves. Hardware such as DAQ card is needed to communicate between software and coupled tank. Because of that, the limitation for this hardware must be considered. The PID algorithm is also needed to develop the GUI for this controller.

There are differences at graph plot between the simulation and implementation results because of the error happen at implementation result due to hardware limitation such as the voltage at capacitive level sensor are not equal with the voltage that set at the coding of the controller. If there is no error, the implementation result should tally as the simulation result.

5.2 Future Recommendation

1. Solenoid valve can be used at coupling point.

2. Real time system can be implemented to get more accurate results.

3. Ultrasonic sensors can be used instead of capacitive probe type to get high accuracy even for 1 mm resolution also.

4. Apart from try and error method to tuning gain for each parameter, PID Controller tuning through other method such as Ziegler Nichols and Cohen Coon tuning formulae etc.

5. Issue of hardware limitation that affected the experiment result. This can be solved by placing The RC circuit can be placed between the DAQ card and coupled tank connection as a filter to get the smooth result.

REFERENCE

[1] Jutarut Chaorai-ngern, Arjin Numsomran, Taweepol Suesut, Thanit Trisuwannawat and Vittaya Tipsuwanporn, " PID Controller Design using Characteristic Ratio Assignment Method for Coupled-Tank Process", Faculty of Engineering, King Mongkuts Institute of Technology Ladkrabang, Bangkok 10520, Thailand, 2005

[2] Muhammad Rehan, Fatima Tahir, Naeem Iqbal and Ghulam.," Modelling, Simulation and Decentralized Control of a Nonlinear Coupled Tank System", Department of Electrical Engineering, PIEAS, Second International Conference on Electrical Engineering University of Engineering and Technology, Lahore (Pakistan), 25-26 March 2008

[3] M. Khalid Khan, Sarah K. Spurgeon, " Robust MIMO water level control in interconnected twin-tanks using second order sliding mode control", Control and Instrumentation Group, Department of Engineering, University of Leicester, Leicester LE1 7RH, UK, 10 February 2005

[4] Liu Jinkun., MATLAB simulation of advanced PID control. Beijing: Electronic Industry Press, 2004

[5] Qiang Xiong a, Wen-Jian Cai a, b,*, Mao-Jun He a, Equivalent transfer function method for PI/PID controller design of MIMO processes, Aug 28, 2012